LIGHTS ON!

ANIMALS THAT GLOW
OCTOPUSES AND SQUID

by Joyce Markovics

CHERRY LAKE PRESS
Ann Arbor, Michigan

CHERRY LAKE PRESS

Published in the United States of America by Cherry Lake Publishing Group
Ann Arbor, Michigan
www.cherrylakepublishing.com

Reading Adviser: Beth Walker Gambro, MS Ed., Reading Consultant, Yorkville, IL
Content Adviser: Steven Haddock, Senior Scientist and Marine Biologist
Book Designer: Ed Morgan

Photo Credits: © Divelvanov/Shutterstock, cover; Margaret McFall-Ngai/Wikimedia Commons, title page; © freepik.com, copyright page and TOC; © Mark Conlin/Alamy Stock Photo, 4; © David Shale/naturepl.com, 5 top; Margaret McFall-Ngai/Wikimedia Commons, 5 bottom; Brooke et al, NOAA-OE, HBOI, 6; © Shpatak/Shutterstock, 7; © David Shale/naturepl.com, 8; © Divelvanov/Shutterstock, 10-11; Wikimedia Commons, 11 top; Wikimedia Commons, 12; Michael Vecchione/NOAA/Wikimedia Commons, 13; Michael Vecchione/NOAA/Wikimedia Commons, 14; Image Courtesy of NOAA Bioluminescence and Vision on the Deep Seafloor 2015, 15; © optionm/Shutterstock, 16; NOAA Office of Ocean Exploration and Research, 2015 Hohonu Moana, 18; © Nature Picture Library/Alamy Stock Photo, 18-19; © rook76stamp/Shutterstock, 19; © David Shale/naturepl.com, 20; © Sönke Johnsen, 21; Kletr/Shutterstock, 22 top left; © Sönke Johnsen, 22 right.

Cherry Lake Press is an imprint of Cherry Lake Publishing Group.

Library of Congress Cataloging-in-Publication Data
Names: Markovics, Joyce L., author.
Title: Octopuses and squid / by Joyce Markovics.
Description: Ann Arbor, Michigan : Cherry Lake Publishing, [2022] | Series: Lights on! animals that glow | Includes bibliographical references and index. | Audience: Grades 4-6
Identifiers: LCCN 2021035285 (print) | LCCN 2021035286 (ebook) | ISBN 9781534199620 (hardcover) | ISBN 9781668900765 (paperback) | ISBN 9781668906521 (ebook) | ISBN 9781668902202 (pdf)
Subjects: LCSH: Octopuses–Juvenile literature. | Squids–Juvenile literature.
Classification: LCC QL430.3.O2 M376 2022 (print) | LCC QL430.3.O2 (ebook) | DDC 594/.56–dc23
LC record available at https://lccn.loc.gov/2021035285
LC ebook record available at https://lccn.loc.gov/2021035286

Printed in the United States of America
Corporate Graphics

CONTENTS

SHINING SUCKERS

In 1997, ocean scientist Edith "Edie" Widder made a twinkling discovery. She was in a deep-sea **submersible** nearly 3,000 feet (914 meters) underwater in the Gulf of Mexico. The water around her was as black as night. Using a light on the sub, she scanned the darkness for life. Edie was on the hunt for bioluminescent animals.

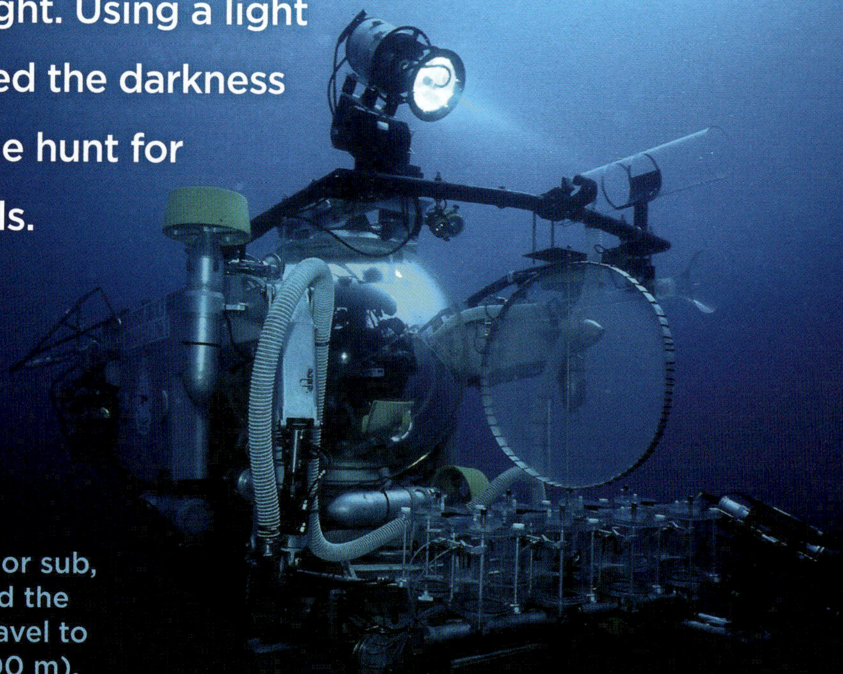

The deep-sea submersible, or sub, Edie was riding in is called the *Johnson Sea Link*. It can travel to depths of 9,843 feet (3,000 m).

Suddenly, something the size of a football caught Edie's eye. It was a puffy octopus the color of a ripe tomato. Edie had never seen anything like it. She sprang into action, knowing the only way to learn more about the creature was to safely catch it!

The red octopus Edie Widder saw

Ocean scientist Edie Widder

Bioluminescence (buy-oh-loo-muh-NES-uhns) means "living light." It's a chemical process that allows animals to make their own light.

Working from the inside the submersible's **hull**, Edie and the pilot used a remote-controlled tool to gently scoop up the octopus. They sealed it inside a water-filled box on the outside of the sub. Then the pilot steered the sub back to the ocean's surface.

The *Johnson Sea Link* on the ocean's surface with a sample in its collection box

Edie climbed out of the sub onto a big ship that served as a floating science lab. In her hand, she held the box containing the live octopus. She and another scientist, Sönke Johnsen, took photos of the newly discovered creature. They closely studied the octopus's body. "You know, those don't look like suckers on the octopus's arms," Sönke said.

An octopus's eight arms are covered in suckers, which look like suction cups.

Octopuses are masters of **camouflage**. They have special skin cells that allow them to change their skin color or texture. This helps them blend into their **habitat**—or even express their mood!

Edie agreed with Sönke. "They look like light **organs**," the two scientists exclaimed. They carried the octopus to another part of the ship and turned off the lights. Before their eyes, the octopus's suckers glowed a bright blue!

The octopus Edie and Sönke discovered is sometimes called the glowing sucker octopus.

The two scientists had just discovered one of only three known bioluminescent octopuses—and the only one with light-up suckers! Because the place where these octopuses and other deep-sea animals live is so dark, they make their own light to survive. In fact, three out of four deep-sea animals are said to use bioluminescence.

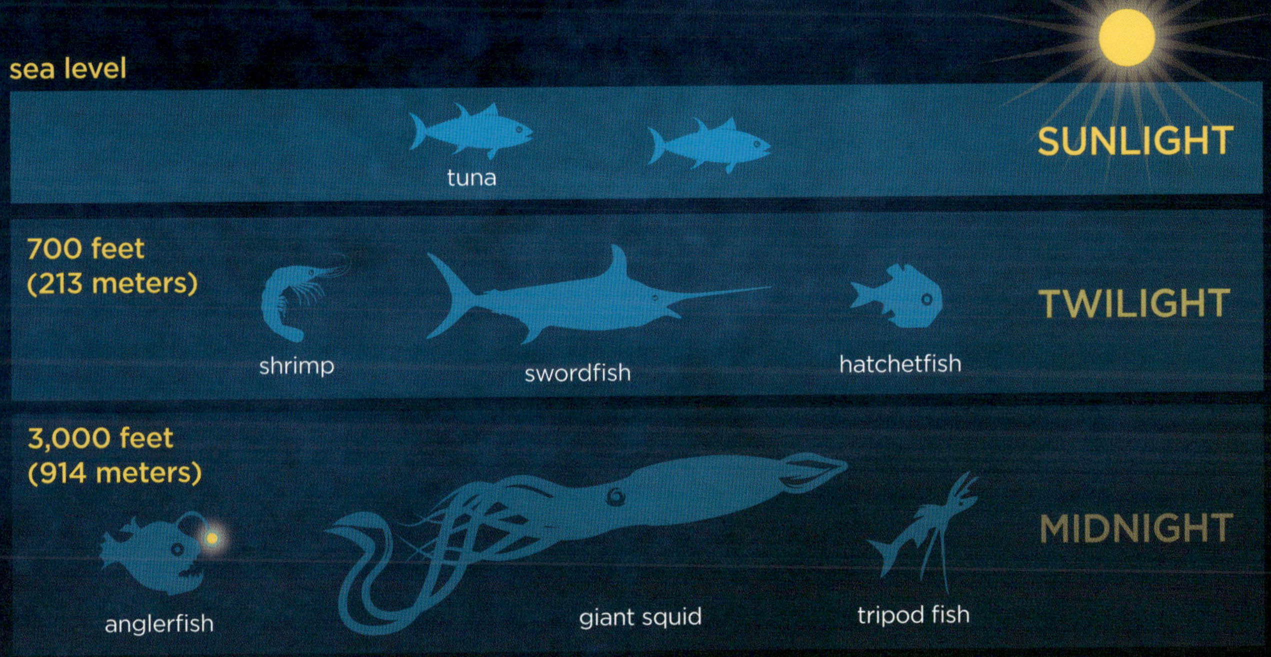

The ocean is divided into different zones based on light and depth. One of the deepest zones, where no light reaches, is called the midnight zone. This is where bioluminescent octopuses make their homes.

SEA LEGS

Octopuses are cephalopods (SUH-fuh-luh-podz), a group of sea animals that also includes squid. The word cephalopod means "head footed." Why? Octopuses and squid have arms or **tentacles** that surround the base of their heads. Most of their organs are in a saclike body above their heads.

Octopuses and squid vary in size from less than 1 inch (2.5 centimeters) across to more than 50 feet (15 m) long. The longest cephalopod is the giant squid. It's almost the length of two school buses!

A dead 30-foot (9-m) giant squid

Octopuses have eight arms and no tentacles. Squid have eight arms and two retractable tentacles. Arms and tentacles are used to feel around, grasp things, and grab prey.

Some octopus and squid **species** make their own light—but how? In their bodies, a chemical called luciferin (loo-SIH-fuh-rin) mixes with oxygen. This process creates energy in the form of light.

This tiny deep-sea *Bolitaena* octopus makes its own light.

A female *Bolitaena* octopus has a yellow bioluminescent ring around her mouth. Scientists think it helps her attract the opposite sex.

While most cephalopods make their own light, some get help from tiny glow-in-the-dark friends—bioluminescent **bacteria**! In exchange for light, the cephalopods provide the bacteria with a safe home inside their bodies. This type of relationship is known as **symbiosis**.

Bioluminescence can help sea animals look invisible, confuse predators, find food, or even attract a mate!

SURPRISING SQUID

The midwater, or eye-flash, squid puts on a stunning glow show. About the length of a human thumb, it's covered with 550 tiny spots. The squid uses these spots, called photophores, to make its own light in the darkest parts of the sea.

Midwater squid live in the Atlantic and Mediterranean Oceans. These sea dwellers often swim to shallower water at night to feed on tiny shrimp.

The midwater squid doesn't use this light to see in the dark, however. It uses light to hide! Squid are a tasty meal for many ocean predators, such as sharks and whales. To stay safe, a squid lights up to match the sunlight or moonlight coming from above it. This is called counter-illumination (ih-loo-muh-NAY-shuhn). From below, the squid is almost impossible to see!

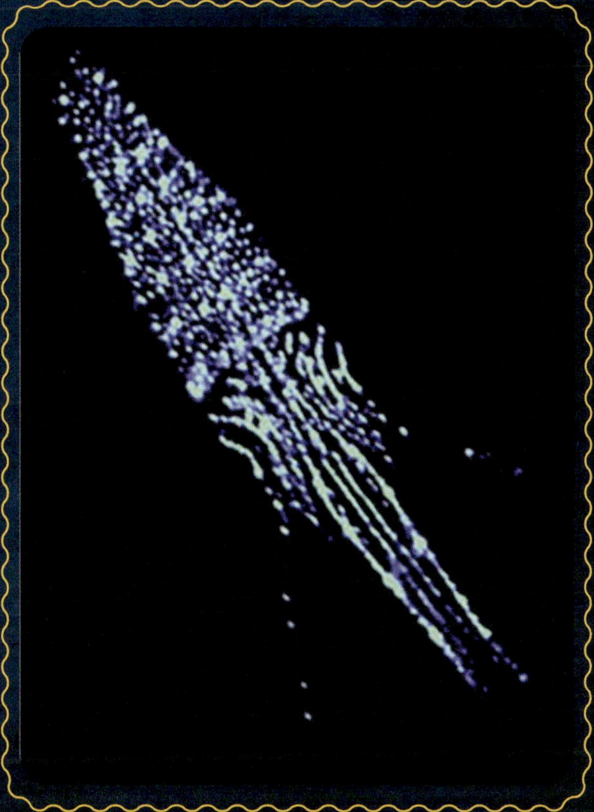

Depending on its environment, a midwater squid can glow green, blue, or a combination of the two.

Scientists have identified many more bioluminescent squid than octopuses. However, there are likely hundreds of deep-sea species yet to be discovered!

Like the midwater squid, the bobtail squid is tiny and glows to hide from enemies. However, this squid shines bright thanks to bioluminescent bacteria! The bacteria live in a special light organ in the squid's body. There, they feed on sugars made by the squid.

The bobtail squid is active at night.

Each day, the bobtail squid **replenishes** its supply of glowing bacteria. It captures a new batch using a net of sticky **mucus**. The bacteria gather in the light organ. At night, the bacteria glow when the squid is hunting for food. As a result, the squid doesn't cast a shadow, keeping it safe from attackers.

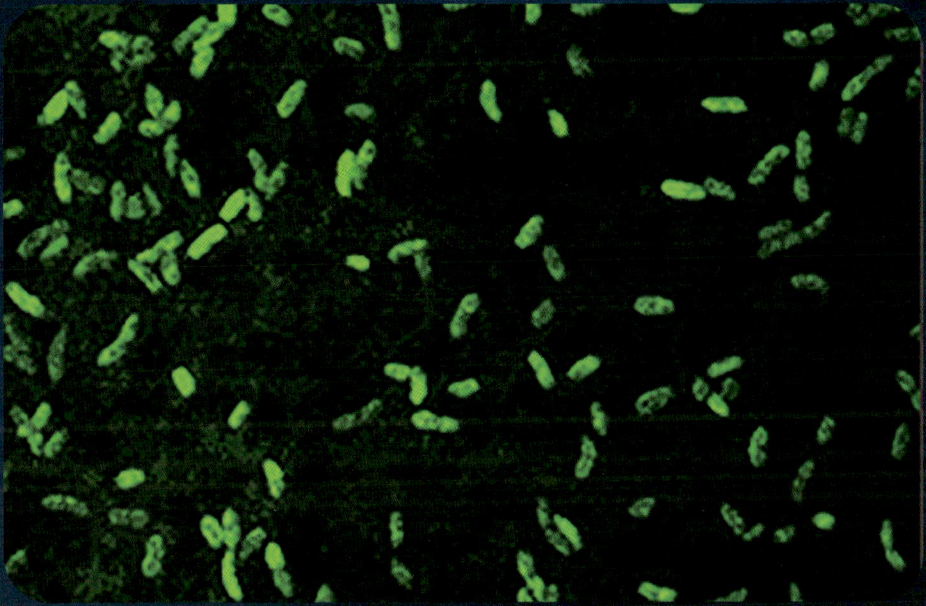

The tiny bioluminescent bacteria partners of the bobtail squid

A cousin of the bobtail squid squirts a cloud of light-up bacteria at predators to confuse them! It's no surprise this species is called the fire shooter squid.

The bobtail squid is no match for the Dana squid—one of the largest cephalopods on Earth! It can weigh as much as a gorilla and grow 7 feet (2 m) long. In 2005, a group of scientists in Japan filmed the animal for the first time in its habitat. That's when they saw the giant squid's thumb-size photophores lighting up like headlights.

The Dana squid

The Dana squid doesn't just light up, it flashes! It can open and close black skin flaps over its photophores, almost like eyelids. Then the squid flashes in different patterns. Scientist think it might do this to blind its prey—or perhaps to attract a mate.

Firefly squid live only for about one year.

Bioluminescent firefly squid also live off the coast of Japan. During mating season, these tiny squid gather in the thousands near the shore, causing the water to glow blue.

AMAZING ADAPTATIONS

Losing an arm might seem like a big deal. It's not if you're a deep-sea squid called *O. deletron*! When under attack, this bioluminescent species will **sever** its own arm. The arm will then wriggle and glow brightly for about 30 seconds. This gory trick distracts hungry predators.

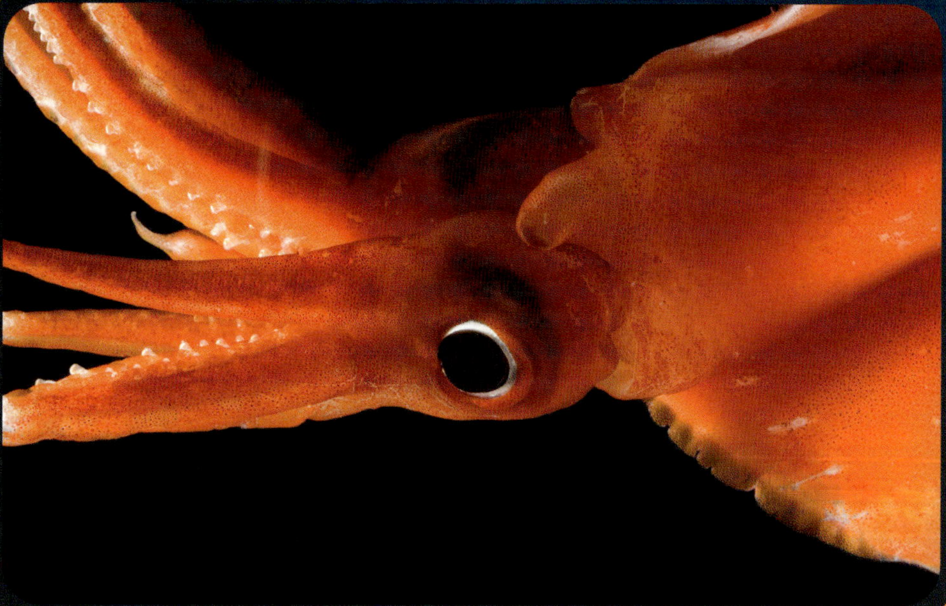

An *O. deletron*, or Octopoteuthis squid

The cockeyed squid is covered in jewel-like photophores. Here it's shown with its big, bulging eye!

The cockeyed, or jewel, squid prefers to keep its limbs. However, it has its own amazing **adaptation**. Its bulging left eye is twice the size of its right one! Scientists believe the big eye helps the squid spot prey that use counter-illumination. In other words, it can spot the shadows of prey above it. How many other incredible deep-sea species are still waiting to be discovered?

Another amazing cephalopod is the vampire squid. It has a cloak-like web that can flip inside out. And it can release a cloud of glowing slime to escape an enemy!

CREATURE FEATURE

BRITTLE STARS

- Like octopuses and squid, brittle stars are multi-armed creatures. However, they're closely related to starfish. Brittle stars crawl across the ocean floor using their five slender, whiplike arms.
- More than 60 kinds of brittle stars are known to be bioluminescent. Most produce green light, while a few make blue light.
- When a brittle star is touched, it produces very bright light. Experts think the brightness is a kind of alarm to drive off an attacker.

GLOSSARY

adaptation (ah-dahp-TAY-shuhn) a change that helps an animal better survive in its environment

bacteria (bak-TIHR-ee-uh) tiny life forms that live everywhere and can be seen only under a microscope

camouflage (KAH-muh-flahzh) coloring or patterns on animals' bodies that match their surroundings

habitat (HAH-buh-tat) a place in nature where animals normally live and are found

hull (HUHL) the frame or body of a ship

mate (MAYT) a male or female partner

mucus (MYOO-kuhss) a sticky liquid made by an animal

organs (OR-guhnz) body parts that do particular jobs

predators (PREH-duh-turz) animals that hunt other animals for food

prey (PRAY) an animal that is hunted by another animal for food

replenishes (ri-PLEH-nish-iz) makes full or complete again

retractable (rih-TRAK-tuh-buhl) something that can be drawn back or inside

sever (SEH-vuhr) to cut off from the body

species (SPEE-sheez) types of animals or plants

submersible (suhb-MUHR-suh-buhl) a special vehicle designed for deep-sea exploration

symbiosis (sim-bee-OH-suhs) when two different organisms live together in a relationship that benefits both

tentacles (TEN-tuh-kuhls) the two long, armlike body parts that a squid uses to grab food

FIND OUT MORE

Books

Davidson, Rose. *Glowing Animals*. Washington, DC: National Geographic, 2019.

Martin, Lee Anne. *Amazing World: Sea Creatures*. Bellevue, WA: Quarto Publishing, 2017.

Regan, Lisa. *Way to Glow! Amazing Creatures That Light Up in the Dark*. New York, NY: Scholastic, 2016.

Websites

Monterey Bay Aquarium Research Institute: Cockeyed Squid
https://www.mbari.org/the-curious-eyes-of-the-cockeyed-squid/

NOVA Science Now: Glowing in the Dark
https://www.pbs.org/wgbh/nova/sciencenow/0305/04-glow-nf.html

Smithsonian: Six Ways to See Ocean Bioluminescence
https://ocean.si.edu/ocean-life/six-ways-see-ocean-ocean-bioluminescence

INDEX

ABOUT THE AUTHOR

Joyce Markovics loves writing about animals that seem too bizarre to be real. The deep sea is a treasure trove of understudied life forms. She hopes this book inspires young readers to take action to protect our oceans and the amazing animals in them. She dedicates this book to KC, an octopus lover, on her fortieth birthday.